# The Good Energy Guide

Published by the Ethical Marketing Group
*www.thegoodshoppingguide.co.uk*

# Foreword

### by the Ethical Marketing Group

Welcome to *The Good Energy Guide* from the Ethical Marketing Group.

We hope you find the following information useful for choosing the best green electricity supplier. Although any green supply is better than dirty old brown, you will see that some suppliers are better and greener than others.

We've also written a section on general energy saving tips that will help too - even if you haven't yet switched to green.

Go on, switch today and become part of the solution – all you need is an old bill, a telephone and about two minutes.

# Good Energy

Welcome to The Good Energy Guide. We hope you find the following information useful for choosing the best green electricity supplier to switch to. Although any green supply is better than dirty old brown, you will see that some renewable suppliers are better and greener than others - so give the good guys your money!
We've also written a section on general energy saving tips that will help you too - even if you haven't yet switched to green.
Go on, switch today and become part of the solution – all you need is an old bill, a telephone and about two minutes!

## GOOD ENERGY IS GREEN...

...Last year the biggest source of carbon dioxide emissions was power stations, accounting for 29 percent of the total. Coal power stations are the least efficient and although the increased popularity of natural gas burning has reduced our potential $CO_2$ emissions slightly, benefits are offset by the continued increase in overall energy usage. Our increasing electricity consumption requires more and more electricity generation, and although consumers' energy efficiency can help reduce this, the only real alternative is to source electricity from renewable resources.

## CONFUSING MARKET

The domestic energy market is confusing enough – a few years ago customers knew that one gas company supplied their gas and nothing else, and another did the same with their electricity. Since 1999 all customers have been able to change their gas or electricity supplier and over 19 million customers have swapped in search of a better deal. Now homeowners have a dazzling array of tariffs and service providers, before even attempting to take the environment into account. Most of the main energy companies provide some kind of green tariff for electricity – the price and coverage depends on the area in which you live but it is generally accepted that green electricity tariffs cost the consumer either about the same or just a few pounds per bill more than conventional tariffs.

Green energy supply has been available in this country to some customers from as far back as 1997. However, it did not truly become an option for the average consumer until the energy market was completely opened up to competition in May 1999. Since then the offerings that are available to us have come a long way.

The green energy revolution has gained

# GOOD HOME

significant support at a commercial level. Large energy users and corporations have taken to green energy supply in a big way. It is not only the case that large 'green'-centric companies such as The Body Shop have a green power supply, but also large institutions such as Oxford University who have 100% of their energy needs provided for by green supply.

Any company, small or large, that claims 'corporate social responsibility' that has not yet switched to a renewable energy supply should think again!

With fourteen green energy tariffs available now in the UK, there is a lot of choice available to the consumer. However, it is not the case that these tariffs all offer the same product.

The most important issue to those on a budget may be the issue of cost. For an average household, as you might expect due to economies of scale, the cost of receiving a green energy supply is fractionally more expensive, the supplement over and above a "regular" tariff is normally in the order of £10 or £20 a year. And changing your methods of payment to either Direct Debit or paying one annual fee can, in most cases, offset the entire extra cost, so there is no reason not to change your supply today.

## Why switch to green energy?

When we read our newspapers and watch the news on TV everyday and see the environmental disasters and freak weather conditions that are attributed to Global Warming, we can see for ourselves the effects of the by-products of our traditional energy production. Electricity production is the single biggest contributor to the emissions that cause climate change. The prime gas responsible for Global Warming or 'The Greenhouse Effect' is Carbon Dioxide or $CO_2$. The burning of Oil, Coal and Gas, otherwise known as Fossil Fuels, in traditional power stations produces a considerable amount of this gas. The UK, with 1% of the world's population, emits 2.3% of the world's total amount of $CO_2$. Not only do these Fossil Fuels contribute to the degradation of the environment, but also they are finite in nature. It is only a matter of time before the planet's supply of these fuels runs out.

One alternative power source to traditional fuel burning stations is Nuclear Power. This is far from being a solution to global pollution though. Although British Nuclear Fuels Limited (BNFL) has been pushing Nuclear Power as the non-polluting solution to climate change, this is certainly not the case. During its lifetime (around 30-40 years) a nuclear reactor can produce radioactive waste that has a 'lifespan' of thousands of years. The waste needs to be disposed of safely, as it is highly dangerous. Although no $CO_2$ is produced there are definitely by-products to the nuclear process that potentially could do serious harm to the environment.

In contrast to these more traditional forms of energy supply, is renewable energy or 'Renewable' Energy. Not only does green energy not directly result in any by-products that may be harmful to our environment, it comes from renewable and everlasting sources such as wind and water. In fact most forms of renewable energy produce no or very little amounts of waste, and therefore have very little impact on the world around us.

# GOOD ENERGY

# 'Last year the biggest source of carbon dioxide emissions was power stations, accounting for 29 percent of the total'.

When you invest in renewable energy supply, you are also supporting the future of the renewable energy industry. By showing the government and mainstream energy suppliers that you wholeheartedly support renewable energy you can help convince them to increase the support they offer to the industry as a whole.

## How do you switch?

The great thing about switching to a green energy tariff is that it's incredibly EASY to do. There is no need to get electricians in, or have anything changed physically with your electricity supply. This is down to the nature of the types of green energy tariff available to the consumer, the Energy-based tariff and the Fund-based tariff and tariffs that offer a combination of the two.

Of the choices available, the Energy-based tariff is the option that actually offers you renewable energy in return for your money. Whilst there is no change in the actual electricity coming down the wires into your home when you subscribe to an Energy-based tariff, a proportion of what you pay will be matched by the equivalent amount of energy being fed into the national grid from renewable sources.

Tariffs such as the one from unit[e], switch at www.unit-e.co.uk, promise to match 100% of the units of electricity you buy from them with an equal amount from renewable sources, at the end of the year.

With Fund-based energy tariffs a proportion of money you pay the supplier is donated into a fund that supports new renewable capacity, green causes or other related initiatives. An independent body, established either by the supplier or a registered charity, normally administers these funds. In some cases the donation made from the consumer is matched in equal amounts by a donation made by the tariff supplier. A Combination tariff is usually some mixture of both Fund-based and Energy-based supply.

Because of the way these tariffs operate, it is extremely easy to switch to a green energy tariff. All you need to do is register your interest with a supplier and they can sign you up over the phone or send you forms to fill out by post. It's also possible to switch your supplier with very little hassle online, at *www.uswitch.com* where you can arrange to pay by Direct Debit, which will also save you money on your bills.

# Change to the leading green electricity product today[1]

**get powered by 100% renewable electricity**
**unit[e] GOOD energy**
www.unit-e.co.uk

## 5 reasons to switch to unit[e]

**1 Rated No. 1 by Friends of the Earth** as the cleanest greenest choice of electricity.

**2 It's easy** You can sign up in less than 10 minutes - call us on 0845 456 1640 or visit www.unit-e.co.uk.

**3 Only 100%** unit[e] is the only UK supplier that supplies _only_ 100% renewable products.

**4 Committed to renewable** unit[e] supports the renewable energy industry by buying electricity from small scale UK generators including renewable electricity generated by individuals and businesses, as well as investing in its own generation.

**5 Know you are making a difference** unit[e] customers can keep up to date with developments at unit[e] and aspects of the renewable electricity industry through the unit[e] newsletter, email newsletter and our website.

**Switching to unit[e] to help keep our world a habitable place is easy**
- Call **0845 456 1640***
- click **www.unit-e.co.uk**

*lines are open 8.30am to 6.30pm Monday to Friday

unit[e], Monkton Park Offices, Monkton Park, Chippenham SN15 1ER

# 'Any company, small or large, that claims 'corporate social responsibility' that has not yet switched to renewable energy supply, should think again!'

## Choosing the best supplier

Since April 1st 2002 Energy suppliers have had to make sure that at least 3% of all energy they provide comes from renewable energy sources. For each unit of renewable energy that they buy they receive a certificate. If companies fail to match their required 3% they may buy certificates from those companies that exceed their minimum.

In order to reach their minimum requirement, large energy suppliers offer a green tariff to customers. In many cases this does not exceed or match the minimum 3% renewable energy that the supplier is required to provide, as demand from traditional tariffs is still considerably greater. These suppliers then have to buy in certificates from smaller niche companies who only offer a green tariff, or their green tariff makes up more than 3% of their total energy supply. If, however, the niche company sells all its certificates other than the 3% it retains to meet its own government targets, it results in a net status quo in the energy market. No extra demand for renewable energy supply is generated, as total demand for renewable energy is matched across the board. Trading of certificates at this level will mean that the net average of renewable energy supply will remain at 3% nationwide. However if those suppliers that produce more than the minimum requirement set aside a further percentage of its certificates, above and beyond the required minimum, refusing to sell them on, additional demand for renewable energy sources is generated. At the moment only unit[e] does this.

When trying to evaluate which tariff is 'better,' it's best to look at what green tariffs are trying to achieve. Ultimately the aim is to increase the amount of renewable energy supply there is in the country. By increasing the influence of renewable energy sources, it is possible to lessen the influence of the environmentally degrading sources, Fossil Fuels and Nuclear Power. It's for this reason that purely Energy-based tariffs are the most positive choice.

## GOOD HOME

# Here are some of the best Green tariffs on offer

### unit[e]
This tariff scored highest in the Friends of the Earth supplier table (reproduced below). Unit[e] is an independent energy provider who only supply energy from renewable sources. It will match 100% of their customers energy use with energy from renewable sources. It buys its energy from small hydroelectric plants and wind-farms. Almost 25% of unit[e]'s existing energy demands are supplied by their wind-farm at Delabole, which was the first ever wind-farm in the UK. In addition to this unit[e] offer to buy back any renewable electricity you might generate yourself for example from solar panels or wind turbines. The main reason why unit[e] tops the FOE suppliers table is that in addition to meeting government targets, they set aside a further 7% of their renewable energy certificates. This helps to generate extra demand for renewable energy sources.

### Ecotricity
Ecotricity is an independent energy supplier that invests in large wind-turbines. It has multiple wind-power generators already operational around the country. At Swaffham in Norfolk it built the country's first multi-megawatt wind-turbine, which alone provides enough energy for 3,000 people. The renewable energy certificates earned by Ecotricity are sold on to help other energy suppliers meet their government targets. The profits earned from the tariff and sale of certificates are then used to build further wind farms and turbines.

### Eco Energy
This is the highest rank combination tariff in the Friends of the Earth tables. It's provided by Northern Ireland Electricity. They let their customer choose whether some or all of their electricity is matched with supplies from green sources. In addition to this Northern Ireland Electricity run the Eco Energy Trust into which a proportion of their income will be deposited. This fund is used to fund community renewable energy projects. As there is no legal minimum requirement for renewable energy production in Northern Ireland, all the energy you buy is 100% green, as there is no need to sell on this demand to help other suppliers reach legal requirement for renewable energy.

## GOOD HOME

# Friends of the Earth Green Energy League Table

| Rank | Product | Supplier |
|---|---|---|
| 1st | **UNITE[E]** | Unit Energy Ltd |
| | Only supplier to exceed legal targets by 7%. Deals in green energy only. Best green deal | |
| 2nd | **Green Energy 100** | Green Energy UK |
| | Energy only. Helps meet legal target. Deals in green energy only. Profits used to invest in green energy. Customer receives shares in the company. | |
| =3rd | **Ecotricity** | Ecotricity |
| | Energy only. Helps meet legal target. Deals in green energy only. profits used to build wind farms. | |
| =3rd | **Eco Energy** | Northern Ireland Electricity |
| | Combination. No legal targets in NI so customers buy 100% of greenness. Plus fund for community renewables projects. | |
| =3rd | **Green Energy 10** | Green Energy UK |
| | Energy only. Helps meet legal target. 10% energy bought is green. Profits used to invest in green energy. Customer receives shares in the company. | |
| =6th | **Green Tariff** | London Electricity Group |
| | Combination. Helps meet legal target. Fund gives £25.60 pa (incl 50% match funding) to community projects plus £64,000 donated at start. | |
| =6th | **Juice** | Innogy group |
| | Energy only. Helps meet legal target. Acts as petition in favour of an Npower off-shore wind farm. | |
| =8th | **Green Energy Offer** | Scottish Power |
| | Combination. Helps meet legal target. Fund gives £21 pa (incl 50% match funding) to community projects and info schemes. | |
| =8th | **RSPB Energy** | Scottish and Southern Energy Plc (SSE) |
| | Combination. Helps meet legal target. Donation made to RSPB (between £5-15 pa) for climate related projects. | |
| 10th | **Green Fund Tariff** | Seeboard Energy Ltd |
| | Pure fund so won't help meet legal targets. Fund gives £30 pa (incl 50% match funding) to community projects but has yet to make an award. | |

This league table was compiled and produced by Friends of the Earth in Spring 2002. It has not yet been matched in terms of complete analysis. For full explanations go to *foe.co.uk* and contact suppliers for latest updates.

To assess the greenness of each of the products FoE have ranked products according to criteria which FoE based on OFGEM's guidelines for green electricity tariffs. See *foe.co.uk* for a full indepth guide

# GOOD ENERGY

## FINAL THOUGHTS: THE FUTURE

Despite the differences in green tariffs available, and the ranking of one above another, switching to any green supply is a positive step to take. It is a win-win situation both for the environment and your peace of mind. Whether or not you choose a Fund-based tariff or an Energy-based tariff, what you are doing when you switch is registering your support for more environmental awareness from the energy suppliers. This will help encourage those suppliers who currently do not offer a green tariff to start one, which is clearly a good thing.

Your vote for cleaner energy supply also has an impact on the future of government policy. For example, by the end of 2004 all energy suppliers will have to disclose the exact sources of their electricity. Green energy supplier, unit[e], have decided to spearhead this disclosure and have shown the market the way forward by making their sources publicly available already. So by supporting green energy suppliers you can also show your support for government reform.

It has never been easier to switch your energy supplier than it is now. All it takes is a simple phone call, or at *www.uswitch.com* and you can start helping to create a cleaner planet. So why wait! (The three box chart below is our own summary of The Ethical Consumer Research Association's research (see opposite page), which includes an examination of the activities of the Company Group as a whole - including the ultimate holding company. For this reason, the ECRA table varies from the Friends of the Earth table. It is up to you to decide which renewable energy company to support - ideally choose a company that score well on both analyses)!

- Ecotricity
- Green Energy 100
- Unit[e]

- Eco Energy
(Northern Ireland Electricity)
- Green Energy
(Scottish Power)
- RSPB Energy

- Green Plan
(Powergen)
- Juice
(N power)
- Green Tariff
(London Electricity)
- Green Fund
(Seeboard)

# GOOD HOME

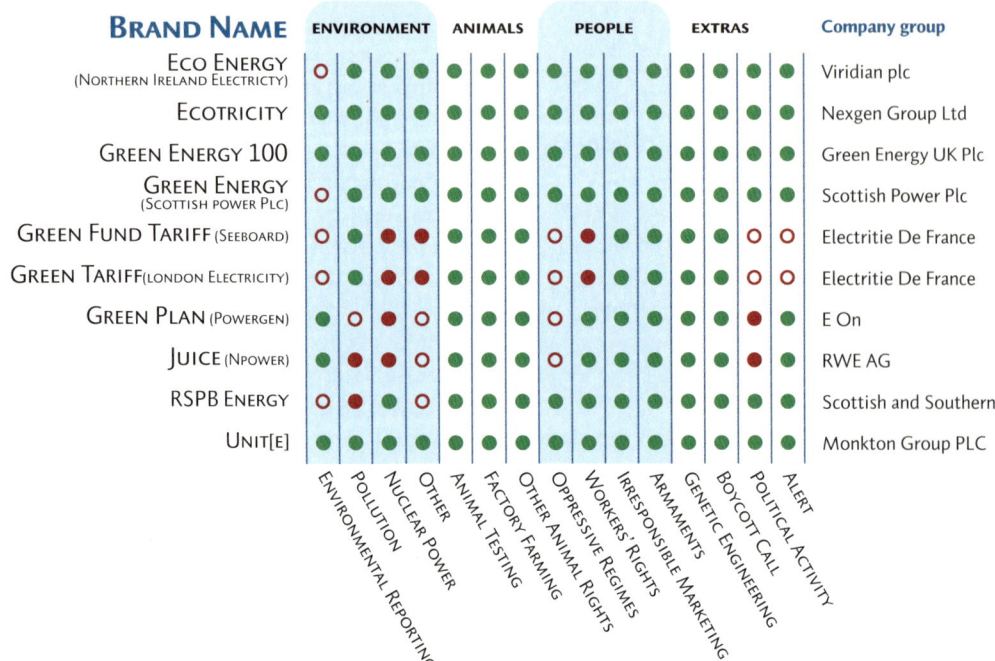

## Key

- 🟢 Top rating (no criticisms found)
- ⭕ Middle rating
- 🔴 Bottom rating
- 🟠 A related company has a bottom rating and the company itself has a middle rating
- ⚪ A related company has a middle rating
- 🟡 A related company has a bottom rating

Source: The Ethical Consumer Association, August 2003. Marks on the tables represent criticism from campaign groups worldwide and are sourced from ECRA's database which contains around 45,000 references on around 25,000 companies.
(🟢) Top rating: a green circle indicates that we have found no criticisms. (⭕) Middle rating: an empty red circle indicates a certain degree of involvement in that category.

(🔴) Bottom rating: A full red circle represents the worst level of involvement in that category.
All marks relate to the company group as a whole, so can include activities of all companies associated with or owned by the ultimate holding company.
For an extended explanation of these symbols and the category definitions go to www.ethicalconsumer.org or ask ECRA for the 'Introduction to Ethical Consumer' pamphlet.

# GOOD ENERGY

## To switch now: Contact

**Unit[e]**
Freepost NAT 4921
Chippenham, Wiltshire. SN15 1BR
Domestic customers Phone: 0845 456 1640
Business customers Phone: 0845 456 1650
www.unit-e.co.uk

**Green Energy UK: Green Energy 10/100**
190 Strand, London. WC2 8JN
Phone: 0845 566 9550
www.greenenergy.uk.com

**Ecotricity**
Axiom House, Station Road, Gloucester. GL5 3AP
Phone: 01453 756 111
www.ecotricity.co.uk

**Northern Ireland Electricity: Eco Energy**
120 Malone Road, Belfast. BT9 5HT
Phone: 08457 455 455
www.nieenergy.co.uk

**London Energy: Green Tariff**
40 Grosvenor Place, Victoria, London. SW1X 7EN
Phone: 0800 096 5060
www.london-energy.com

**Innogy Group: Juice**
NPower Centre, Oak House, Bridgewater Road, Warnden, Worcester. WR4 9FP
Phone: 0800 316 2610
www.npower.com

**Scottish Power: Green Energy Offer**
1 Atlantic Quay, Glasgow. G2 8SP
Phone: 0845 270 6543
www.scottishpower.co.uk

**Scottish and Southern Energy plc: RSPB Energy**
Southern Electric, PO Box 6009, Basingstoke. RG21 8ZD
Phone: 0845 7444 555
www.southern-electric.co.uk

**Seeboard energy: Green Fund Tariff**
Phone: 0800 096 9696
www.seeboardenergy.com

Also see u-switch.co.uk

# www.THEGOODSHOPPINGGUIDE.co.uk

# The Good Energy Guide to Energy Efficiency

A standard three bedroom detached house, without any forms of insulation, can cost up to £500 a year to heat. With proper energy efficiency measure taken it is entirely possible to halve this cost. It's not only through heating that your energy efficiency in the home can be improved. Changes to your lighting and household appliances can also help reduce the amount of energy you consume.

## Heating

During the cold winter months we all rely on our heating to keep us warm and cosy. However, having an energy inefficient heating system can result in you spending more than you need to on your heating costs. Here are some tips on how you can improve your heating efficiency.

Make sure you have an effective method of heating control. Boilers are unable to tell when you want heat or hot water without a form of heating control. If some form of heating control is installed you can regulate when and where you need heat. Controlling heat efficiently around the house can save you up to 17% on your heating costs.

If your boiler is more the 15 years old you should think about replacing it. New energy efficient Condensing Boilers could save you up to 32% on your fuel bills. Even without upgrading to a Condensing Boiler, modern, more efficient boilers can still save you up to 20%. In addition to this your local Council may be able to provide a grant to help you out.

If you live in a small property, you could also consider using energy efficient convection heaters or gas heaters to heat your property rather than relying on central heating.

## Lighting

In the average home you can expect your lighting costs to account for 10-15% of your electricity bill. With lighting accountable for such a sizeable percentage of your costs, it seems only sensible to invest in ways in which you can improve efficiency around the house. With energy saving lightbulbs now readily available, here's some further information:

| Ordinary Bulbs | Energy Saving |
|---|---|
| 25W | 6W |
| 40W | 8-11W |
| 60W | 13-18W |
| 100W | 20-25W |

# GOOD HOME

Energy saving lightbulbs only use a quarter of the energy that standard bulbs do. For this reason they are available in much lower wattages (see table). However the light from an energy saving bulb is often radiated differently to a conventional one, so you may need to choose an equivalent higher wattage bulb than you are used to achieve the same lighting effect.

At the moment energy saving lightbulbs tend to be more expensive to buy than conventional ones, at around £5 for a 20W bulb. However the cost benefit makes up for this extra initial outlay. For every conventional bulb you replace with an energy saving one it could save you up to £10 a year on your electricity bill, making back the £5 spent on the bulb and leaving you with an extra £5 in your pocket.

To complement energy saving bulbs, you could consider having energy saving fittings in which to place them. These are little transformers that fit into the base of the bulb which regulate the amount of energy that is fed into it. For the few milliseconds it takes for a bulb to light the transformer provides a surge of energy. Once a bulb is lit it requires far less power to stay alight, so the fitting maintains the electricity flow into the bulb at a very low level.

## HOUSEHOLD APPLIANCES

No matter how well you feel your household appliances are running and how few problems they have given you, they could still be extremely energy hungry and inefficient. As a general rule, the older your appliance the more it is going to cost to run. For this reason, where possible it is best to buy your fridges, cookers, dishwashers and washing machines brand new as they will be the most energy efficient. The saving you make on a second-hand purchase will soon be outweighed by the extra cost it takes to run the appliance. When buying new appliances look out for the Energy Efficiency Recommended logo. To find out more about which appliances currently available are listed as Energy Efficient, go to *www.saveenergy.co.uk* and browse the extensive database of Energy Efficient household appliances.

## INSULATION

Bad insulation in the home can result in a considerable heat loss. Most heat is lost through the walls and the loft space. Fully insulating these spaces can help reduce the amount of heat lost in the home by more than 50%. The walls alone can be responsible for up to 35% of the total heat wastage in the home.

Badly insulated walls can be one of the major sources of heat loss in the home. They could be costing you anywhere up to £200 extra per year. For this reason insulating the walls of your home is one of the most efficient ways to make a saving on your heating bills. If you want to find out what you can do about adding insulation to your walls, the first step is to identify what kind of walls you have in your home. Most houses built after 1930 have cavity walls. To identify whether you have cavity walls you can check by measuring their thickness at a door or window. They are normally around 30cm thick, this is comprised of an inner and outer layer, and in-between them is a small air gap. To fill your wall with insulation, small holes are drilled into the

outer or inner layer and insulation material is injected into the air gap. This work has to be carried out by a professional, and will be guaranteed for 25 years by the CIGA or Cavity Insulation Guarantee Agency. The cost of the work should be recovered within five years in the savings you make on your heating costs. There are also grants and offers available to help cover the cost of the work.

As air gets hotter it becomes less dense, and as a result of this rises above cold air, which is denser. This is the reason why it is key to make sure any heat lost through the roof is minimised. Most houses have some space under the roof, normally the loft. Insulating the loft properly can save around 25% on your heating costs. You can insulate your loft easily yourself, and requires no professional work to be done. By simply adding a 250mm (10-inch) thick layer of insulation the job is done. The material that you need to insulate the roof can easily be picked up at a local DIY store or builder's merchants.

Drafts coming through the edge of the skirting board or up through the cracks in the floor can make a room feel cold and unwelcoming. Sealing up these cracks with a regular tube sealant can save you up to £10 in your heating bills. To make your floors warmer and to stop the chilly drafts coming up through them you could invest in some under floor insulation, which can help save a further £25. Remember, if you fit the insulation yourself, not to block any air bricks on the outside wall. These help maintain adequate ventilation under the floor, and without this it's likely that the floorboards will start to rot.

Heat that escapes through the space under your doors or windows also accounts for a considerable amount of heat lost in the home, as much as 20%. Draft excluders come in many different materials, from brushes to rubber strips. Without double-glazing these can be a cheap and easy way to prevent heat escaping from your home. Do remember that in some rooms ventilation is very important, especially if they have solid fuel burners, gas fires or boilers within.

Badly insulated hot water pipes and boilers can result in 75% more energy use than those that are fully insulated. British Standard boiler "jackets" can be found at all good DIY stores and are easy to fit. The saving you make on your water heating bills means the cost can be recouped within a year. If you already have insulation on your boiler check that its at least 75mm (3 inches) thick. If it isn't it could be a good idea to replace it with a new one to make yours as energy efficient as it can be.

For further information on improving insulation you can get in touch with your local Energy Efficiency Advice Centre. If you don't know where this is you can phone freephone 0800 512 012 or search on the Energy Savings Trust website at *www.saveenergy.com*.

## GLAZING

Double-glazing your windows is an ideal way to reduce heat loss in the home by up to 20%. Whilst it is an expensive option, it should definitely be considered if you are planning on renovating your window frames. Not only does double (or even triple-) glazing help prevent heat loss but can also stop condensation and reduce

# GOOD ENERGY

noise levels of sounds from outside. If you are on a tight budget you can always fit secondary glazing, which is less expensive than fitting brand new double-glazing and can still result in annual savings of around £30.

## Quick Tips to Improve your Energy Efficiency Today!

● If you are too warm at home, turn down your thermostat by 1°C. This could save you up to 10% on your heating bill. If you are planning to go away over the winter for any extended period of time, turn the thermostat down to a low level. You can turn it down as far as you want, but be sure to leave it high enough so the house doesn't freeze. Your total saving could be as much as £30 a year.

● There is no need to have the hot water come out of your taps at scalding temperatures. For most people a setting of 60°C/140°F on their cylinder thermostat will be more than enough for taking baths and washing-up. Doing this can save you as much as £10 a year.

● Never leave the taps running and the plughole unblocked. If you are washing up or using hot water, try not to do it with the plughole open. The cost for hot water can soon mount up and leaving the plughole open can flush money away with the wastewater.

● Always close your curtains in the evening. Your curtains are a valuable form of insulation. If you close your curtains you can stop extra heat escaping out through the window into the cold night air.

● Try not to use electric lights when there is a good source of natural light available.

Open your curtains or blinds fully rather than switch on an electric light. If you do use an electric light make sure you remember to switch it off when you leave the room.

● Electrical devices such as Television and Computers consume almost as much electricity in their standby mode as when switched on. Try to switch off all devices of this nature if you can. Obviously if this will have an effect on the appliance's memory settings then leaving it on standby can be unavoidable' so check the appliance's manual before you switch it off.

● Defrosting your fridge and freezer can help it run more efficiently; try to do this as often as possible. Also try not to leave the fridge or freezer door open for more than a few seconds as the cold air will escape, meaning the appliance will have to work harder to cool the air inside down again when you do close the door.

● It's important to try to make sure you run a full load in your washing machine and tumble dryer. If this is impossible, use the economy wash settings or run at a low heat. Modern washing powders will work just as effectively at 40°C as at 60°C. These rules can apply to dishwashers too; try to run a full load every time and use the lowest temperature setting available.

● When cooking try to use the best pot or pan available for the job, and match this with the right cooking ring. Ideally the base of the pot should just cover the edges of the ring. If you are using a gas hob the flames should only heat the bottom of the pot, any flames that rise up the sides of the pot will be wasted heat.

● When boiling water in a kettle, there is no need to fill it all the way to the top if you

'Fossil Fuels contribute to the degradation of the environment, and are finite in nature. It is only a matter of time before the planet's supply of these fuels runs out.'

are not going to use all the water. Fill the kettle with enough water to cover the element, but not more than you plan to use.
- A tap left dripping for a day can waste as much water as it would take to run a good sized bath. This is needless waste, especially if the water is hot. Make sure you firmly close all taps when you have finished with them.
- If you are used to taking baths, consider switching to a shower. An ordinary shower uses less than a half of the water that a bath does. You can easily buy devices that convert your bath taps into a shower.

Buy copies of this book and others at *www.thegoodshoppingguide.co.uk*